The Royal Military Canal

A BRIEF HISTORY

GEOFF HUTCHINSON

Contents

Other titles by the same author:

FULLER OF SUSSEX — A Georgian Squire
THE MARY STANFORD DISASTER — The story of a lifeboat, November 15 1928
GREY OWL — The incredible story of Archie Belaney, 1888-1938
BAIRD — The story of John Logie Baird, 1888-1946
HUMGRUFFIN — A story of country folk . . . and giants
THE LOVERS' SEAT — The history and the love story of 1786. Fact and fiction
KIPLING — Rudyard Kipling, 1865-1936. An introduction
AN INTRODUCTION TO HASTINGS & ST LEONARDS
AN INTRODUCTION TO BEXHILL-ON-SEA
AN INTRODUCTION TO BATTLE, RYE AND THE VILLAGES
MARTELLO TOWERS — A brief history

Acknowledgements

I wish to acknowledge the following for the invaluable assistance given to me during the preparation of this booklet: Alan and Sheila Jones for their continued encouragement and support; Mr Tony Lackner of New Hall, Dymchurch; The National Rivers Authority; Mr C Lock for sowing the seed; Keith Fitz-Hugh; Colin Spray of Dataset; Mr R Sheridan for loan of pictures; the ladies of Battle Library; Jan Roadnight; and all the friendly and helpful people I met while walking and cycling the towpaths of the Royal Military Canal in the summer of 1994. This booklet is intended purely as an introduction and a sister publication to *Martello Towers — A brief history*, published in 1993. For those interested in further detailed reading, I recommend *The Canal Builders* by Anthony Burton and P A L Vine's excellent book *The Royal Military Canal* which, although published some years ago, still remains the 'bible' for this fascinating subject.

Cover picture: Hythe, Kent, by J M W Turner, *circa 1823.*
Reproduced by kind permission of the Guildhall Museum and Art Gallery, Corporation of London

---

First published 1995

ISBN 0 9519936 3 1

Design, artwork and typesetting by the author
Text set in 10/12pt Century Book
Printed by M & W Morgan, Red Lake Terrace, Ore, Hastings

Invasion looms

In the late 18th century, following the upheaval of bloody revolution, the French embarked on expansionist policies. Led by the military genius of a young Corsican officer named Napoleon Bonaparte, the all conquering French armies rampaged a path across Europe. In 1793 France declared war on England.

After nine years of fighting a treaty was signed between the two countries which brought peace. However, this respite was to be short-lived and in May 1803 hostilities resumed. England now found herself in crisis and not for the first time in her history — or the last — faced the real threat of invasion.

During the late summer of 1803, public opinion and enthusiasm to strengthen the country's coastal defences were strong and military experts reasoned that if the French were to attack, they would take the shortest route and attempt to land their troops on the flat open beaches which lay between the natural cliff defences at Folkestone in the east and Hastings in the west. Civilians were encouraged to take up arms as troops of volunteers were formed to supplement the regular forces on the coast.

However, the Government, led by the weak and indecisive Prime Minister, Henry Addington, seemed incapable of making a structured plan to resist the French challenge and was loath to provide the necessary materials and weapons. While Addington dithered for the best part of a year, preparations for war were being made across the Channel, near Boulogne, as Bonaparte, now self-proclaimed Emperor of France, amassed a mighty army of over 130,000 troops and assembled a flotilla of 2000 boats to make the crossing. He knew his greatest obstacle was the British Navy, which prevented him gaining access to the Channel, but by 1804 his confidence was high as he waited, hawk-like, for the moment to strike.

Fortunately, the British Government was to change that year, when William Pitt the Younger, who had spent the previous few years as Lord Warden of the Cinque Ports, was recalled to office to begin his second term as Prime Minister. Pitt's enthusiasm marked a sharp contrast to Addington's hesitancy and although not a well man, he soon made the defence of the coast a top priority.

In early September 1804, Pitt called a meeting at the New Hall, Dymchurch, on the Romney Marsh, to discuss the various methods of defence. Present at this gathering were some of the most prominent military dignitaries of the day.

Earlier in the year a scheme had been put forward to erect a chain of defensive forts (Martello towers) along the coast. A second idea was to flood the whole of the Romney Marsh by opening a series of sluices. The third plan was to construct a canal to serve as a defensive barrier. This latter suggestion was to become one of the most incredible — and eccentric — feats of military engineering ever embarked on in Britain.

The Royal Military Canal

BY COURTESY OF HASTINGS HERITAGE LTD

Napoleon Bonaparte — confident of success
Below: a floating fortress, part of his intended invasion fleet

Planning

In 1804 the hierarchy of military command was as follows: overall responsibility for defence of the realm was in the hands of the Commander-in-Chief of Armed Forces, the sometimes maligned Duke of York; under his command and responsible for the co- ordinated defence of the country, was General Sir David Dundas, Commander of the Southern District and beneath him Sir James Pulteney, responsible for the Sussex district and Sir John Moore for Kent. These much respected soldiers advised on ways of making the coast secure against enemy attack and it was the Commander-in-Chief's duty to gain Government approval.

However, the initial suggestion for a canal came from Lt-Col John Brown, Assistant Quartermaster-General and Commandant of the Royal Staff Corps (now the Royal Engineers). Brown had spent a considerable time exploring the feasibility of the flooding scheme, which he eventually rejected as unworkable. He reasoned that at least ten days warning would be needed to implement the plan and even then it was doubtful if the whole marsh could be flooded completely at a neap tide. He also highlighted the embarrassment, chaos and huge expense which would be created in the event of a false alarm.

Brown's plan for a canal was submitted to his superior, Dundas, in September 1804. Brown suggested that a vast ditch be built from Shorncliffe, near Folkestone to the River Rother near Rye, a distance of 19 miles and that a road be built alongside. The road would be screened from the enemy by a parapet built from the soil dug from the canal. As well as being a line of defence the canal would provide the facility to move troops either by barge or by road to the threatened points. Brown suggested that the canal be 62ft wide at the surface, 44ft wide at the bottom and nine feet deep. The adjoining parapet would be 35ft wide and the road 30ft. The canal would be built in angled

John Rennie

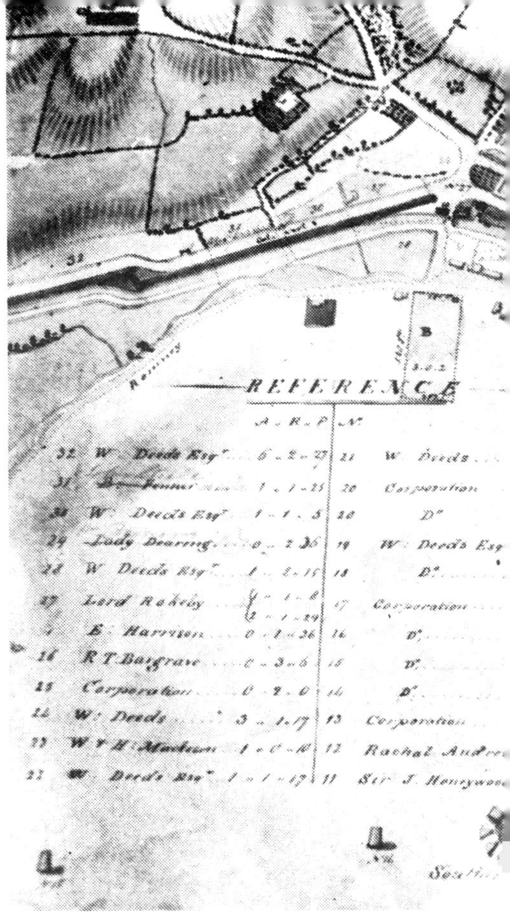

Right: Part of the original plans for the Royal Military Canal. This section shows its course through the town of Hythe and lists the names of the landowners along its path

sections, with enfilading cannon fire covering each stretch and would follow the path of the old Saxon shoreline across the Romney Marsh skirting the villages of West Hythe, Bonnington, Bilsington, Ruckinge, Ham Street, Warehorne, Kenardington, Appledore and on to Iden where the Rivers Rother and Brede would form the defence work as far as Winchelsea.

The Duke of York showed great enthusiasm for the project and preliminary planning was undertaken with amazing speed. The scheme was discussed by Pitt and the Duke on September 26, forwarded to the Minister of War, Lord Camden the following day and in marked contrast to the uncertainty of his predecessor Addington, Pitt ordered that work should begin immediately.

The celebrated John Rennie, who had been responsible for the construction of London and Waterloo Bridges, was appointed consultant engineer and Brown was appointed military director. Rennie was

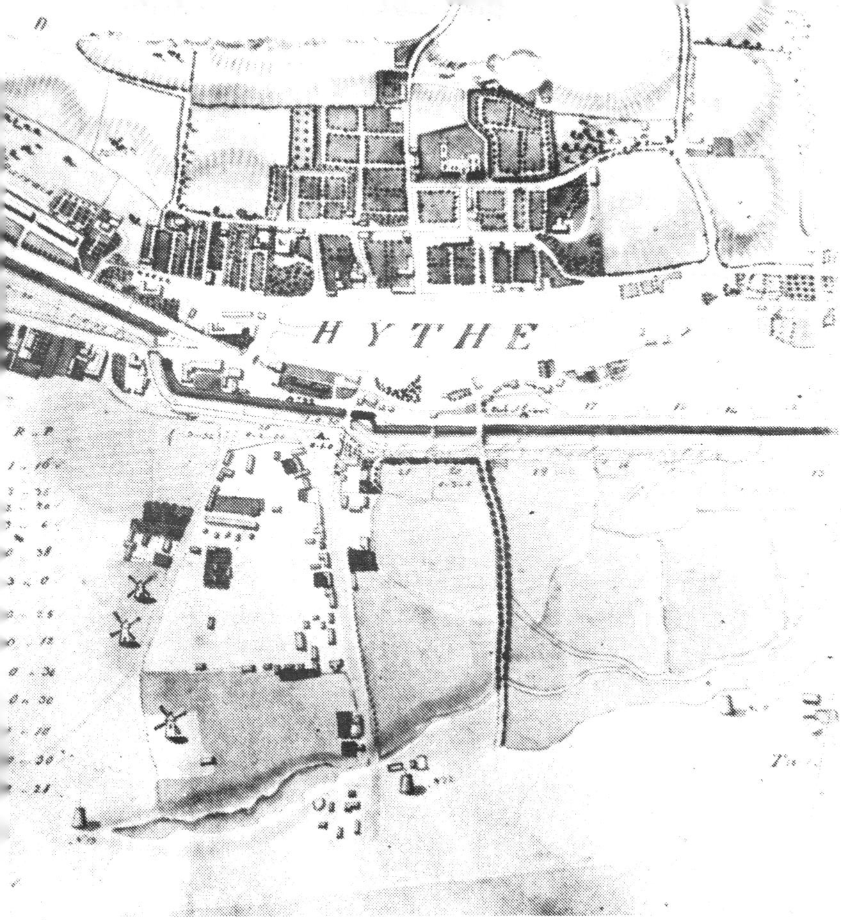

BY COURTESY OF THE CORPORATION OF ROMNEY MARSH

responsible to the Quartermaster-General Major General Brownrigg, the officer in charge of the direction and management of the works, and one week later prepared a survey report on the canal. In it he praised Brown's original plan and suggested a few refinements; a sluice at Shorncliffe, culverts to take flood water under the canal and a lock near Rye. It was also suggested that the canal be extended from the River Rother through to Pett Level, three miles east of Hastings, but this section would not be as wide or as deep. The total length of the defence work would be 28 miles, of which 22½ miles had to be dug.

It was now down to Pitt to convince the local landowners of the advantages of the scheme. At a further meeting at the New Hall, Dymchurch on October 24 he explained that the new canal would not only help to defend their country but would serve as a major drainage system for the winter and a reservoir for the summer. Pitt's personal involvement and subtle persuasion was most instrumental in smoothing out early problems and land was soon made available to him. He was popular with the locals of Romney Marsh, who fondly referred to the proposed canal as 'Mr Pitt's Ditch'.

The speed with which Pitt had organised this mammoth project was quite remarkable. In little over a month from the initial decision to go ahead with the scheme, preparations were complete and the time had come for work to commence.

William Pitt the Younger, Promoter of the Royal Military Canal and below, the New Hall at Dymchurch in Kent, venue for several important meetings concerning the southern coastal defences

BY COURTESY OF THE CORPORATION OF ROMNEY MARSH

Building

On October 30 1804, the first soil was dug at Shorncliffe. It was intended that the canal would be ready for use by June of the following year, but it soon became clear that this estimate was wildly optimistic as harsh winter weather and severe flooding caused much delay. Frustrated by the lack of progress, Rennie blamed the incompetence and greed of the contractors, who were accused of overcharging and inadequate supervision. By May 1805, the scheme seemed on the brink of disaster. Only six miles had been completed, work had stopped, the contractors dismissed and Rennie had resigned.

The project was then put under the control of the Quartermaster-General's department and Lt-Col Brown was placed in command. He took personal charge of the eastern end and Lt-Col Nicolay supervised the section between Winchelsea and Pett Level. As a result the work ran more smoothly with a combination of civilian and cheaper military labour, supervised by the Royal Staff Corps, which had moved its headquarters permanently to Hythe. The civilian labour was used to dig the canal, while the military built the ramparts and turfed the banks. Initially hand pumps were employed day and night to keep the trench from filling with water, but the problem of flooding was a constant drawback to progress. Eventually powerful steam-driven engines were used to drive the pumps in an effort to clear the water faster, but the planned depth of nine feet could still not be achieved in many places along the canal's route.

During the summer of 1805, work also began on the south coast Martello Towers, under the supervision of Brigadier General William Twiss, and the area was soon turned into a massive building site. This sudden influx of workers had a considerable impact on the local community. By August the number of men working on the canal alone was over 1500. The work was hard and often performed in inclement

BY COURTESY OF HASTINGS MUSEUM AND ART GALLERY

weather. Digging was accomplished by hand, using only picks and shovels and the soil was carried away in wheelbarrows and loaded onto horse-drawn carts. Once the canal was dug it had to be lined with a thick mixture of wet clay — a process known as puddling — which stopped water seeping through the sides.

These tasks were carried out by navigators — better known as navvies — who travelled all over the country for work. They often led a tough and unruly life. When the soil was soft an experienced navvy was capable of shifting 12 cubic yards of earth a day, which represents a trench three feet wide, three feet deep and 36 feet long. However, it is doubtful if this was ever achieved on the Royal Military Canal, due to the hardness of the ground in winter months and a figure of around two cubic yards would have been more realistic. They were paid around five shillings a week for their labours, which they spent freely and although the area welcomed the boost to its economy, it was not so keen on the rise in crime, drunkenness and other associated social problems.

Much of Rennie's early thinking on the canal had been directed towards controlling its water supply; to stop excess water entering in winter and spring and to keep it topped up in summer. This was

This engraving from Turner's Rye, Sussex, *illustrates the hazards experienced during the building of the canal. In the foreground the construction of a junction between the canal and the river Brede has been taking place behind a temporary dam. The dam has been overwhelmed by the spring tide which has sent the navvies fleeing or clinging to the timber shuttering for safety. The town of Rye can be seen in the distance and Camber Castle is to the right with the sea beyond*

achieved by placing drains at the ends of the canal and by diverting smaller streams into it in summer.

As work progressed, The Royal Waggon Train (the predecessor of The Royal Army Service Corps) dragged barges on to the newly finished parts of the canal to provide shingle for the adjoining road and later trees were planted along the edge, the roots of which gave added support to the parapet.

Although the first sections of the canal at its most easterly end were cut to the planned width of 62ft, increasing problems experienced by the builders as it snaked its way across Romney Marsh must have caused them to compromise as work progressed. At most points on the Marsh the canal is just half its projected width.

In August 1806 the canal was open from Shorncliffe to the River Rother and the Duke of York inspected the work from a small boat towed by relays of horses taking 2¾ hours at a speed of 7mph. But there were still problems. Several incidents of storm damage meant that sections of the canal had to be rebuilt and this was compounded by a serious fire at the regimental stores in Hythe which caused delays.

In 1807 there were many accusations of profiteering and the Government decided to put the canal on a more legal footing when it passed an Act to establish its maintenance. The Royal Military Canal Act of August 13 1807 appointed an illustrious board of commissioners to administer the canal. They included the Prime Minister, the Speaker of the House of Commons, the Chancellor of the Exchequer, the Commander-in-Chief of the Armed Forces and Master-General of the Ordnance and the Quartermaster-General.

In September of 1808 Iden lock was completed. This connected the canal to the Rother Navigation and Rye Harbour and in effect turned Romney Marsh into an island. The canal was now capable of conveying large numbers of soldiers between Sussex and Kent.

It was not until spring 1809 that the canal was effectually completed, although the defensive cannons were not put in position until the summer of 1812. Lt-Col Brown was honoured by the commissioners who paid him £3000 in return for his efforts. He was also made director of the canal and given a house at Hythe.

By the time the Royal Military Canal was fully ready for use, the threat of invasion had long past. It was never to be used for its intended defensive purpose. Napoleon's hopes of invasion had been shattered in 1805, following the defeat of his naval fleet at the Battle of Trafalgar. He withdrew his troops from the French coast and concentrated his hostile intentions on the Austrians.

The final cost of the Royal Military Canal was £234,310, a staggering amount in Georgian England and with the threat gone and no invader to repel, the voices of the cynics and doubters were soon to be heard loud and clear from all sections of English society.

Later history

The critical analysis of the usefulness of the canal as a defensive barrier was to become an embarrassment to the Government. Many considered it to be a white elephant of the greatest proportions and a disastrous waste of public money. Some sceptics reasoned that there had never really been a serious threat of invasion and even if it had occurred, it was most unlikely that the canal would have provided Napoleon's troops with too many problems. As the radical William Cobbett remarked after taking one of his *Rural Rides* through Kent, it was most improbable that the French 'were to be kept back by a canal, made by Pitt, thirty feet wide at the most!' Cobbett was not alone in his criticism. Many blamed Pitt for being responsible for an enormous and costly folly. The criticism would not concern William Pitt, a man who had devoted his whole life to his country. He had died in 1806, tired, sick and lonely, worn out by the rigours of war.

In desperation, the Government tried to find ways of recouping some of the cash expended on the canal and in 1807 opened it to navigation. The Royal Military Canal Act of 1807 enabled the commissioners to collect tolls for the transportation of produce and goods. The commissioners owned a few barges of their own, towed by horses of the Royal Waggon Train and these were soon put into service to transport goods.

In April 1810 the canal was opened to the public and tolls were also collected for use of the military road between Rye and Iden and Rye and Winchelsea. A regular barge service was run between Hythe and Rye, which took around four hours to complete. Despite all efforts to utilise the canal, traffic was never heavy and the Government was unable to recoup its expenditure on the building and maintenance costs.

But the canal had brought some benefits to the villages of the

The swing bridge which spanned the Rye Navigation and carried the Ashford to Hastings Railway, which opened in 1851

Romney Marsh. They now had better means of communication by road and water and the canal had greatly increased the quality of farming land.

In 1837 the responsibility for the running and maintenance of the canal was transferred to the Board of Ordnance and was looked after by an officer and 60 troops. However, this number of men was gradually reduced and by the end of 1841 there were none.

The decline in traffic using the canal was accelerated by the opening of the railway line between Ashford and Hastings in 1851 and financially the canal now had to rely solely on renting land to local farmers for sheep grazing.

Two attempts were made to purchase the canal from the Government during the 1840s; the first by the Hastings, Rye and Tenterden

COURTESY OF HASTINGS MUSEUM AND ART GALLERY

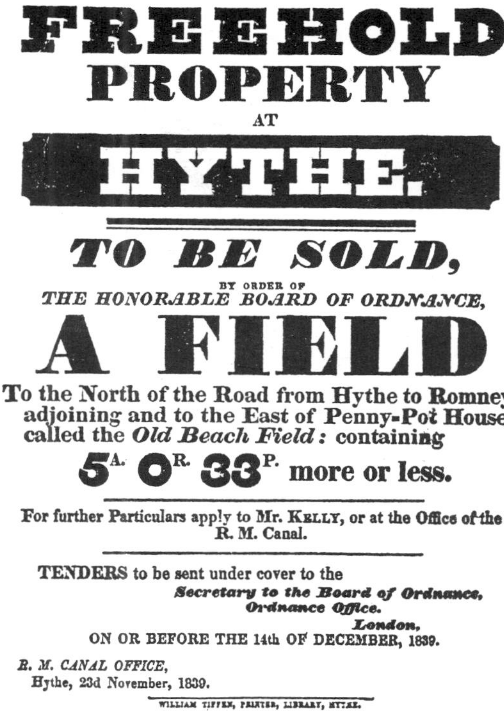

FREEHOLD PROPERTY AT HYTHE.

TO BE SOLD,

BY ORDER OF
THE HONORABLE BOARD OF ORDNANCE,

A FIELD

To the North of the Road from Hythe to Romney adjoining and to the East of Penny-Pot House called the *Old Beach Field:* containing

5A. 0R. 33P. more or less.

For further Particulars apply to Mr. KELLY, or at the Office of the R. M. Canal.

TENDERS to be sent under cover to the *Secretary to the Board of Ordnance, Ordnance Office. London,* ON OR BEFORE THE 14th OF DECEMBER, 1839.

R. M. CANAL OFFICE,
Hythe, 23d November, 1839.

WILLIAM TIFFEN, PRINTER, LIBRARY, HYTHE.

BY COURTESY OF THE CORPORATION OF ROMNEY MARSH

A Notice of Sale issued by the Board of Ordnance in 1839

Railway Company, who wished to convert it into a railway line, a measure defeated in Parliament; the second by the Lords of Romney Marsh, which fell through because of lack of finance.

The Royal Military Canal was never again really taken seriously as a defensive work, although the Duke of Wellington still considered it useful in the 1840s and at the outbreak of the Crimean War in 1855, the canal was taken over by the War Department, but was still only regarded as a modest hurdle to any invader.

During the 1860s traffic still continued at a trickle and the Government decided to take steps to divest its interest in the canal. Eventually the War Department agreed to lease the stretch between Iden and Shorncliffe to the Lords of the Romney Marsh for 999 years at an annual rent of one shilling and the town of Hythe purchased the stretch through the town for conversion into ornamental waters. In 1874 the War Department sold the canal west of Rye to four private owners. By now the canal trade had all but gone. The last toll collected at Iden Lock was taken on December 15, 1909.

The canal and its adjoining road had just one more link to times of war when, in 1935 it was requisitioned by the War Department and in 1940 pill-boxes were placed along its banks, many of which are still there. Later, anti-aircraft guns were set up along the canal to bring down flying bombs on their way to London.

Right: Several pill-boxes from World War II are still standing by the canal. Below: Pleasure boating at Hythe in 1913

The canal today

Today the Royal Military Canal is protected as an ancient monument and is maintained along its whole length by the National Rivers Authority. As in the past, the canal still plays a vital role in the drainage of the Romney Marsh by preventing flooding of arable land in winter and providing water for agriculture in summer. The canal also supports an important coarse fishery and is a haven for wildlife of all descriptions.

The eastern end of the canal is still owned by Hythe Council and is a popular venue for boating in the summer months. Every two years a

colourful Venetian Festival is held at Hythe, which attracts thousands of visitors. The western end from Appledore to Iden Lock is under the control of the National Rivers Authority and in 1935 the section between Appledore and Hamstreet was bought by Dorothy Johnston, who promptly presented it to the National Trust.

The canal is a popular venue for fishing and water sports such as canoeing, and much of the towpath is enjoyed by walkers. The Royal Military Canal Path is waymarked and signed with distinctive blue and yellow emblems.

Summary

Walking the peaceful towpaths of the Royal Military Canal today, it seems hard to believe that it was built as an implement of war in Georgian times. True, the odd pillbox from a later conflict may appear, but it swiftly fades into insignificance. Today the canal is a place of peace, where anglers fish the calm water, swans and other birdlife meander gently across its width and the beauty of its plant life in summer months provides a stunning spectacle. It is a place to enjoy — and reflect on days gone by.

The Royal Military Canal, built by the determination of one man, William Pitt the Younger, who with perhaps over-zealous enthusiasm, persuaded the country that a modest canal could stop a rampaging army. If the initial reasons for its creation seem illogical and the fact that in historical terms it must be regarded as a military folly, the Royal Military Canal would still play an important role in future years. Today it is an integral and vital part of the beautiful Romney Marsh and at the same time serves as yet another reminder of the eccentricities of the colourful days of Georgian England.